STEAM IN THE KITCHEN

I0811827

# ADDING AND SUBTRACTING IN THE KITCHEN

THEIA LAKE

PowerKiDS press

Published in 2024 by The Rosen Publishing Group, Inc.
2544 Clinton Street, Buffalo, NY 14224

Copyright © 2024 by The Rosen Publishing Group, Inc.

All rights reserved. No part of this book may be reproduced in any form without permission in writing from the publisher, except by a reviewer.

First Edition

Editor: Theresa Emminizer
Book Design: Rachel Rising

Photo Credits: Cover, p. 1 VGstockstudio/Shutterstock.com; pp. 4,6,8,10,12,14,16,18,20 porcelaniq/Shutterstock.com; p. 5 pics five/Shutterstock.com; p. 7 Ground Picture/Shutterstock.com; p. 9 Madaland2/Shutterstock.com; p. 11 Niradj/Shutterstock.com; p. 13 RozenskiP/Shutterstock.com; p. 15 Prostock-studio/Shutterstock.com; p. 17 SewCreamStudio/Shutterstock.com; p. 19 Dmytro Zinkevych/Shutterestock.com; p. 21 antoniodiaz/Shutterstock.com.

Cataloging-in-Publication Data

Names: Lake, Theia.
Title: Adding and subtracting in the kitchen / Theia Lake.
Description: New York : Powerkids Press, 2024. | Series: STEAM in the kitchen | Includes glossary and index.
Identifiers: ISBN 9781499443639 (pbk.) | ISBN 9781499443646 (library bound) | ISBN 9781499443653 (ebook)
Subjects: LCSH: Addition–Juvenile literature. | Subtraction–Juvenile literature. | Kitchens–Juvenile literature.
Classification: LCC EQA115.L35 2024 | DDC 513.2'11-dc23

Manufactured in the United States of America

Some of the images in this book illustrate individuals who are models. The depictions do not imply actual situations or events.

CPSIA Compliance Information: Batch #CWPK24. For Further Information contact Rosen Publishing at 1-800-237-9932.

Find us on

# CONTENTS

# Math in Real Life

When you think of math, you may think of sitting in a classroom. You might picture writing out numbers on paper. Did you know that you can also practice math in the kitchen? Cooking and baking are fun ways to do math in a hands-on way!

## Let's Make a Recipe!

There are lots of ways to practice adding and subtracting in the kitchen. Looking at a **recipe** is a good place to start! Let's say you want to make chocolate chip cookies. Your recipe shows a list of **ingredients**. Gather them together and place them on a table.

## Count It Up

Now count your ingredients! How many are there altogether? You can also count how many of a certain kind of ingredient you have. Flour, sugar, and baking soda are dry ingredients. Eggs, butter, and milk are wet ingredients. How many dry ingredients do you have? How many wet?

- flour
- sugar
- baking soda
- eggs
- butter
- milk
- chocolate chips
- walnuts

## Subtract It Out

Are there any ingredients in your recipe that you don't want to use? Maybe your recipe calls for nuts. You may not like nuts, so you want to take them out. Subtract that ingredient from your recipe. How many ingredients do you have now?

8 (all ingredients)
– 1 (1 ingredient: nuts)

7 (ingredients left)

# Add It Together

Perhaps your recipe makes 12 cookies, but you want to make 24. To **double** your recipe, you'll need to add! The 12-cookie recipe calls for one cup of brown sugar. How much brown sugar will you need to add to double the recipe? Add another cup of brown sugar!

$$\begin{array}{r} 1 \text{ cup} \\ +\,1 \text{ cup} \\ \hline 2 \text{ cups} \end{array}$$

## Mix It and Make It

Mix the dry and wet ingredients apart (except for the sugar with the butter), and then together! As you put all your ingredients into one bowl, you can practice adding again. One teaspoon of baking soda plus one teaspoon of salt equals two teaspoons. Watch as everything comes together and turns into **dough**!

## Spoon It Out

You can also practice adding and subtracting as you spoon out the dough. How many cookies can fit on your pan? You might fit three rows of three **(3 + 3 + 3 = 9)**. Or you could make four rows of three **(3 + 3 + 3 + 3 = 12)**. You might need to subtract a row if making bigger cookies **(12 – 3 = 9)**.

## Put It in the Oven

Have a grown-up help you heat the oven to 375°F (191°C). Then set a timer. Bake your cookies for 10 minutes. Once baked, let the cookies cool for 5 minutes. How many minutes is that altogether?

|   | **10 minutes** |
|---|---|
| + | **5 minutes** |
|   | **15 minutes** |

## Let's Eat!

You've practiced adding and subtracting. You've made a tasty treat too! Now it's time to eat. You made 12 cookies. How many are left if you eat one?

$$\begin{array}{r} 12 \text{ cookies} \\ -\ 1 \text{ cookie} \\ \hline 11 \text{ cookies} \end{array}$$

# GLOSSARY

**double:** To cause something to be two times as great or in amount.

**dough:** A mixture of flour and other ingredients that is baked to make cookies, bread, and other baked foods.

**ingredient:** One of several parts that come together to make something.

**recipe:** A list of ingredients and steps to make food.

# FOR MORE INFORMATION

## BOOKS

Askew, Mike. *Let's Add & Subtract*. New York, NY: PowerKids Press, 2022.

Virr, Paul, and Lisa Regan. *Adding Puzzles*. New York, NY: Windmill Books, 2020.

## WEBSITES

**National Geographic Kids: Fun Food Facts**
*kids.nationalgeographic.com/chew-on-this*
Find more fun recipes and interesting food facts here!

**PBS Kids: Math Games**
*pbskids.org/games/math/*
Practice your adding and subtracting skills with these fun games!

Publisher's note to educators and parents: Our editors have carefully reviewed these websites to ensure that they are suitable for students. Many websites change frequently, however, and we cannot guarantee that a site's future contents will continue to meet our high standards of quality and educational value. Be advised that students should be closely supervised whenever they access the internet.

# INDEX